I SPY

At the PARK

by Spencer Brinker

Consultant:
Beth Gambro
Reading Specialist
Yorkville, Illinois

Contents

BEARPORT
PUBLISHING

New York, New York

At the Park

Where am I?
I am at the park!

What do I spy?

I spy a tree.

It is tall.

I spy a dog.

It is big.
Woof!

I spy a swing.

It is red.

I spy a slide.

It is yellow.

I spy a bench.

It is green.

I spy children, too!

They are racing.

Key Words

bench

dog

slide

swing

tree

Index

About the Author

Spencer Brinker lives and works in New York City. In such a big city, you can spy almost anything.

Teaching Tips

Before Reading

- ✔ Guide readers on a "picture walk" through the text by asking them to name the things shown.
- ✔ Discuss book structure by showing children where text will appear consistently on pages.
- ✔ Highlight the supportive pattern of the book. Note the consistent number of sentences found on each page.

During Reading

- ✔ Encourage readers to "read with your finger" and point to each word as it is read. Stop periodically to ask children to point to a specific word in the text.
- ✔ Reading strategies: When encountering unknown words, prompt readers with encouraging cues such as:
    - **Does that word look like a word you already know?**
    - **Check the picture.**

After Reading

- ✔ Write the key words on index cards.
    - **Have readers match them to pictures in the book.**
    - **Have children sort words by category (words that begin with the letter *s*, for example).**
- ✔ Ask readers to identify their favorite page in the book. Have them read that page aloud.
- ✔ Ask children to write their own sentences. Encourage them to use the same pattern found in the book as a model for their writing

Credits: Cover, © Martin Wahlborg/iStock, © photka/Shutterstock, and © Greenseas/Shutterstock; 2–3, © 1000 Words/Shutterstock; 4–5, © Artorn Thongtukit/Shutterstock; 6–7, © gsagi/iStock; 8–9, © Lisyl/Shutterstock; 10–11, © pmmart/Shutterstock; 12–13, © Arporn/Shutterstock; 14–15, © bowdenimages/iStock; 16T (L to R), © Arporn/Shutterstock and © gsagi/iStock; 16B (L to R), © pmmart/Shutterstock, © Lisyl/Shutterstock, and © Artorn Thongtukit/Shutterstock.

Publisher: Kenn Goin **Senior Editor**: Joyce Tavolacci **Creative Director**: Spencer Brinker **Photo Researcher**: Thomas Persano

Library of Congress Cataloging-in-Publication Data in process at time of publication (2019)
Library of Congress Control Number: 2018048658
ISBN-13: 978-1-64280-221-4 (library binding) | ISBN-13: 978-1-64280-394-5 (paperback)

10 9 8 7 6 5 4 3 2 1